T0200245

Series 117

This is a Ladybird Expert book, one of a series of titles for an adult readership. Written by some of the leading lights and outstanding communicators in their fields and published by one of the most trusted and well-loved names in books, the Ladybird Expert series provides clear, accessible and authoritative introductions, informed by expert opinion, to key subjects drawn from science, history and culture.

Every effort has been made to ensure images are correctly attributed, however if any omission or error has been made please notify the Publisher for correction in future editions.

MICHAEL JOSEPH

UK | USA | Canada | Ireland | Australia
India | New Zealand | South Africa

Michael Joseph is part of the Penguin Random House group of companies
whose addresses can be found at global.penguinrandomhouse.com

 Penguin
Random House
UK

First published 2018

002

Printed in Italy by L.E.G.O. S.p.A.

A CIP catalogue record for this book is available from the British Library
ISBN: 978–0–718–18829–0

www.greenpenguin.co.uk

 MIX
Paper from
responsible sources
FSC® C018179

Penguin Random House is committed to a
sustainable future for our business, our readers
and our planet. This book is made from Forest
Stewardship Council® certified paper.

Bubbles

Helen Czerski

with illustrations by
Chris Moore

Ladybird Books Ltd, London

What is a bubble?

Bubbles are beautiful, ephemeral, fun, fragile, jolly and slightly unpredictable. We're all familiar with them, but we don't often ask what they actually *are*.

A bubble is a pocket of gas enclosed in liquid. It sounds simple, and exists because states of matter have boundaries. A gas and a liquid can't merge without becoming purely a gas or purely a liquid. Packaging a blob of gas within the liquid is an efficient way to arrange a mixture.

Humans have loved bubbles for centuries. The ancient Sumerians may have had a word for bubble: *gakkul*, used mostly in the context of beer (and often celebrated for generating happiness).

Bubbles took off in public life with the Victorians and their fascination with soap and cleanliness. Soap consumption in the UK increased from 25,000 tons in 1801 to 100,000 tons in 1861, and as the germs were scrubbed away bubbles floated into everyone's lives, ephemeral symbols of the new world being built by the Industrial Revolution.

Many great scientists in the Western world – Lord Rayleigh, Isaac Newton, Robert Hooke, Agnes Pockles (see opposite, clockwise from top right) and more – studied bubbles seriously. They recognized that bubbles could tell us a lot about the nature of the physical world, and they poked, prodded and listened to find out what it was. In the years since, we've learned that this bulbous arrangement of liquid and gas does things that neither could do by itself.

This book has a message: never underestimate a bubble.

Soap

We all encounter our first bubble very early in life, and it's almost always a soap bubble. They're easy to make and fun to play with, but they've got such a delicate structure that it's amazing they can exist at all. A soap bubble is a pocket of air trapped inside a very thin film of water. Soap molecules coat the inner and outer surfaces of the water so that it doesn't touch the air at any point. This double coating keeps the water layer stable even though it's only about 0.001 mm thick.

Soap bubbles are usually spherical because this shape can contain a fixed amount of gas with the minimum surface area. But the real source of their beauty is their iridescence. The thickness of the soap film is so close to the wavelength of light that when light bounces between the walls some colours are reinforced and others cancel themselves out. The colour you see depends on the angle you look from and the thickness of the soap film at that point.

The thin water layer gradually drains downwards because of gravity (some evaporates from the surface). Glycerol slows down the drainage, which is why it's commonly added to bubble solution. Putting dyes in the soap solution doesn't change the colour of bubbles because there's too little water within the soap film to absorb light. But Japanese astronaut Naoko Yamazaki blew coloured bubbles on the International Space Station. In zero gravity, the water layer is thicker and doesn't drain downwards, so bubbles really can have a single colour.

Soap bubbles, while pretty, are only the beginning. The real world of bubbles is underwater, where a bubble is more robust, and much more interesting.

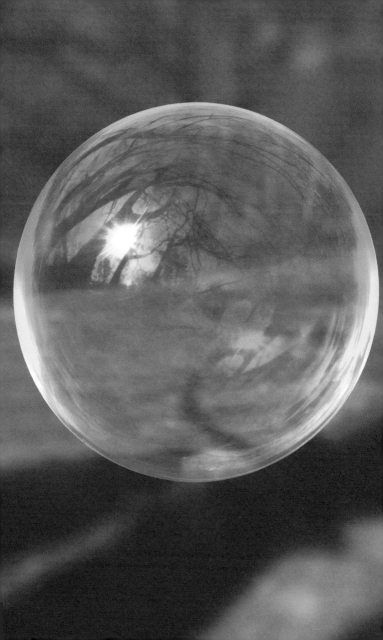

Bubble shape

Underwater bubbles are harder to watch than soap bubbles, but there's much more to see. The most obvious difference is their shape.

A perfectly spherical underwater bubble is a rare beast. Bubble shape is determined by a hierarchy of invisible forces in close competition – the bubble surface is constantly resculpted as these forces jostle for position.

The simplest bubble is a sphere, because it's possible to have a nice tidy balance when there are only two forces in the battle: the gas pressure pushing outwards and the surface tension (the tendency of a water surface to act like an elastic sheet), which acts to reduce the surface area, squeezing the surface inwards. As the volume of the squeezed gas decreases, its pressure increases until a truce is reached – a perfect sphere.

But the only place you'll see this is in zero gravity. Down here on Earth, gravity pulls the water downwards, forcing bubbles to rise. And that brings in a third player – the surrounding liquid pushing on the bubble surface.

Large rising bubbles are flattened into a 'spherical cap' shape, and the bigger the bubble, the more umbrella-like its shape becomes. Turbulent water is full of swirls and mini-currents which can stretch and squish bubbles into lots of weird shapes, and even break them in two. Those forces from the liquid are constantly competing with the surface tension and the bubble's inner pressure.

Almost all bubble shapes can be made by distorting a 'normal' spherical bubble. But there are exceptions, and nature has an expert at producing them: the dolphin.

Dolphins and toroidal bubbles

If you're a dolphin, a bubble is a great toy. And many species of dolphin and whale are so good at playing with bubbles that they've learned how to make a very unusual shape: a ring bubble or 'toroid'. These bubbles are doughnut-shaped – a tube stretched around in a circle – and they're always travelling. They're the aquatic equivalent of a smoke ring.

The reason that these toroidal bubbles can hold their strange shape is that they're really just the core of a spinning doughnut of water. The easiest way to make them is to blow a normal bubble but to add a pulse of pressure at the end, so that a jet of water pierces the middle of the bubble at the last moment. This jet of water will then flow around the outer surface of the bubble ring and back up through the centre, swirling round and round to form a vortex that's curved around in a circle. The ring travels forward, spinning all the time, and it becomes wider and thinner as it goes. It's tricky to get this right in the laboratory, but the dolphins are superstars at it. Both wild and captive animals have been seen making toroidal bubbles, and they often bat the rings to break them up or swim right through their new hoops.

Human swimmers and divers can also make bubble rings underwater, with a bit of practice. But the dolphins definitely beat us hands down at this game.

Antibubbles

An even stranger type of bubble was observed in 1932 by the Rev. William Hughes, a schoolmaster in Southampton. He was dripping soapy water on to a flat water surface, to study the drops that sometimes just sat on top of the water for a few seconds, kept separate by a thin film of air. But occasionally the water surface formed a pit when the drop fell in and then closed behind it. The drop, surrounded by a spherical shell of air, would drift downwards in the water, lasting for up to a minute before its fragile air shell burst.

This is an antibubble, the inverse of a soap bubble. The shell of air is extremely thin – usually 0.01–0.1 mm. Just as with a soap bubble, you sometimes see interference fringes, decorating the antibubble with pink and green stripes. The antibubble is buoyant, but rises very slowly because there's so little air. That gives it a very distinctive perfectly spherical shape.

Just like a soap bubble, an antibubble will burst when its shell gets too thin. In a soap bubble, the water shell gradually drains downwards, so bursting starts from the top, where it's thinnest. But it's the opposite in an antibubble. The thin shell of air drains upwards, and bursting usually starts at the bottom.

Antibubbles are odd, and possibly the strangest bubble shape that exists. But throughout most of history, people haven't spent too much time worrying about what bubbles look like. They've been more concerned with what they do. And the first observation is that a gas bubble in liquid is always rising.

Champagne engines

Bubbles rise because the gas inside is less dense than the surrounding liquid. But a bubble doesn't just rise *through* the liquid – it drags some of that liquid along with it. And that matters, especially in one very famous bubbly drink.

There is evidence that fermented drinks were made in China as long ago as 6500 BC, and if consumed before the gas had escaped, these could have been some of the earliest bubbly beverages. Today, the most iconic bubbly drink is champagne, and the bubbles aren't just there for decoration.

A classic champagne flute has completely smooth sides, but the centre of the bottom of the glass is often etched. These tiny scratches are the perfect place for dissolved carbon dioxide to escape from the drink and form bubbles. The bubbles rise in a thin column, but instead of slipping through the liquid, they drag some of the surrounding champagne up with them. When the bubbles get to the top of the glass, they stop at the surface, but the liquid has to flow outwards and then back down the sides to the bottom. So the bubbles are driving an engine, keeping the champagne circulating. The taller the glass, the faster the engine. The liquid at the surface is continually refreshed, helping flavour molecules to hop from the drink into the air and then float up our noses. Some flavour molecules also collect on the bubble surface, and the bubble spits them upwards when it bursts.

It's easy to appreciate this in a small glass, but these same processes also happen in Earth's vast oceans.

The rising bubbles drive an invisible engine in a champagne glass.

Bubbles (or not) in the Bermuda Triangle

There is a persistent myth that plumes of bubbles from methane seeps could be responsible for the mysterious loss of ships in the Bermuda Triangle. The idea is that a mudslide or sudden fracture at the seabed causes the rapid release of a huge volume of methane gas, and that when this gas reaches the surface the water is temporarily too bubbly to support a ship and so any nearby ship sinks.

We can dismiss this as a myth for two reasons.

The first is that the number of ships lost in the Bermuda Triangle is exactly what you'd expect considering the huge number of ships that pass through that area. There is no mystery to explain.

The second reason is that a deep release of gas is very unlikely to sink a ship. This is because, just as in the glass of champagne, the rising bubbles will drag the surrounding water upwards with them, but gas rising from great depth will form a very rapid plume. The water dragged upwards will erupt above the surface, forming a temporary dome of foam. Calculations and experiments with scale models show that the upward momentum of this water will shove a ship up and outwards, more than compensating for the loss of buoyancy due to the bubbly water. The ship might tip slightly if the plume centre is off to one side, but this is exactly the sort of thing that ships are designed to cope with.

Bubbles won't cause a ship to go down. But it is possible for a bubble itself to move downwards, most famously in a glass of Guinness.

Up or down?

All bubbles rise, but they don't all rise at the same speed. Every buoyant bubble must push water out of the way in order to move upwards. The resistance to the bubble's movement is proportional to its surface area, but the buoyancy is proportional to its volume. So big bubbles experience much less resistance relative to their buoyancy, and rise more quickly. You can see this if you empty a cup of water into a deep tank – big bubbles whoosh quickly up to the surface, while tiny ones lag behind.

It's often observed that the bubbles in a pint of Guinness go down, not up. Most of the dissolved gas in Guinness is nitrogen, which is added after the fermentation process. The bubbles filled with this gas are tiny, perhaps just 0.1 mm across (this is why the foam is so creamy). When the drink is poured, lots of bubbles grow at the bottom of the glass and rise upwards, dragging the liquid in the centre up with them. Because of the shape of the glass, this liquid is replaced by the downward flow of a thin layer around the sides. But this outer layer still contains many tiny bubbles which can only rise very slowly. The outer layer carries them downwards more quickly than they rise upwards. So to the outside eye, it looks as though they're sinking.

You can't usually tell what's inside a bubble just from looking at it. But, like the nitrogen bubbles in Guinness, you can sometimes get a clue from how they form and what they're doing.

The liquid close to the sides of a glass of Guinness is moving downwards, carrying bubbles with it. The bubbles move downwards overall, because the liquid is flowing down faster than the bubbles are rising upwards through it.

Vapour bubbles

You might expect that the gas in a bubble and the liquid around it must be made of different molecules. But they can be made from the same stuff, and the most common example of this is boiling water.

As you start to heat water in a pan, tiny bubbles form around the edges. These are made from dissolved air (mostly oxygen and nitrogen), which escapes from the water as it heats up. The water isn't yet boiling, but the heat does remove these gases from the water.

The water at the bottom of the pan warms up first and then rises, allowing cooler water from above to take its place. This process of convection mixes the water, and the temperature of the whole pan rises at a similar rate. But soon the temperature at the bottom of the pan reaches the point where this warmer liquid has enough energy to turn into a gas (100°C at sea level). A bubble grows at the base of the pan, filled with newly gaseous water molecules. When the bubble becomes big enough, it floats away from the base. But it soon reaches cooler water which takes its energy away, condensing the gaseous water molecules back into liquid. The bubble collapses with a thump before it reaches the surface. Only when the whole pan of water is at the boiling temperature can the bubbles get through all the way to the top. This is why water sounds loudest just before it boils.

But that thump from a vapour bubble collapsing can get far louder than this, and in the claws of one specialist, it's associated with a knock-out blow.

Snapping shrimp

Snapping shrimp are very small and very loud, with a weapon that can knock their prey for six. Each shrimp has one oversized claw, rounded and hardened like a boxing glove. When the claw snaps shut, the water inside gets squirted out through a narrow orifice, generating a mini vortex ring with a powerful water jet barrelling through its centre. That jet can knock out a small fish or another crustacean, which probably never knew that the snapping shrimp was there, and which subsequently goes by the name of 'dinner'.

The jet comes with a bubble-based side effect: the snap that gives the shrimp its name. This bubble is filled with water vapour, but instead of coming from an increase in temperature (as in the kettle), it forms because of a drop in pressure, and this is called a cavitation bubble. As the vortex ring forms, there's a very sudden drop in pressure at its core, ripping the liquid apart. Water can't stretch, so the enormous force pulls a cavity open, and this immediately fills with gaseous water molecules. It's dramatic but short-lived – the vast external pressure pushes back within a fraction of a millisecond, compressing the gas inside the bubble and heating it to around 5000°C. For a fleeting 10 nanoseconds it's so hot that it also emits light. The whip-crack sound comes from the bubble's collapse, a reminder that even a tiny bubble should never be underestimated.

But you don't need to be a shrimp to generate a cavitation bubble. Humans can do it too.

Knuckles

Some people crack their knuckles. This may not make them popular with the people around them, but there is no evidence that it causes any harm. The space inside a finger joint contains synovial fluid, and when someone pulls on that finger, the gap between the finger bones increases. In knuckle crackers, the gap doesn't widen until the pressure drops enough for cavitation bubbles to form. It's the same mechanism as for the shrimp. The gap between the bones then springs open, because the bubbles can't withstand the tension. This time, it's the rapid bubble formation that causes the crack sound. It isn't possible to crack the same knuckle again until the bubble has completely redissolved, which takes perhaps 20–30 minutes.

When he was a child, Donald Unger's mother and several aunts told him that he should not crack his knuckles because it would give him arthritis. Donald did not take their word for it. For the next 50 years, as he was growing up and throughout his career as a medical doctor, he cracked the knuckles on his left hand twice a day but not on his right. At the end of this period, he did not have arthritis in either hand. He wrote a letter in a medical journal about this long-term study, contributing to the growing body of evidence that his mother and aunts were wrong. He was awarded the IgNobel Prize for this work in 2009.

Self-experimentation of this kind is not generally recommended.

Pings and pops

Everyday bubbles can produce sound too. Underwater bubbles are always beautiful to look at, but their real character only comes out when you listen to them. If you have noticed the sound of water being poured into a glass, admired the quiet babbling of a stream or even eavesdropped on your tummy gurgles, you've been listening to bubbles. Bubbles often ring like a bell at the moment of their formation, because at the precise instant they pinch off they briefly have a very strange shape for a bubble: they're pointy. Surface tension rapidly pulls the point inwards, but this acts like a hammer hitting the bell, so the bubble sounds an almost pure note. Just like real bells, a bigger bubble will ring with a deeper note. That means that if you listen carefully you can hear how big a bubble is.

To make sure they're in tune, orchestras play 'concert A', which has a frequency of 440 Hz. A bubble making that note would be a big one: with a radius of 7.4 mm if it was spherical. When you halve the radius, you double the frequency, and for everyday bubbles, the frequency (in Hertz) multiplied by the radius (in metres) is around 3.26.

The Japanese have made a musical instrument based on the sound made by bubbles, called a *suikinkutsu*. A bubble is made by a single drop falling on to a water surface, and the sound is shaped by the ceramic chamber it sits in.

But bubbles don't just make their own sound – they also respond to the sound waves going past them.

A ping is heard at the moment that a new bubble pinches off.

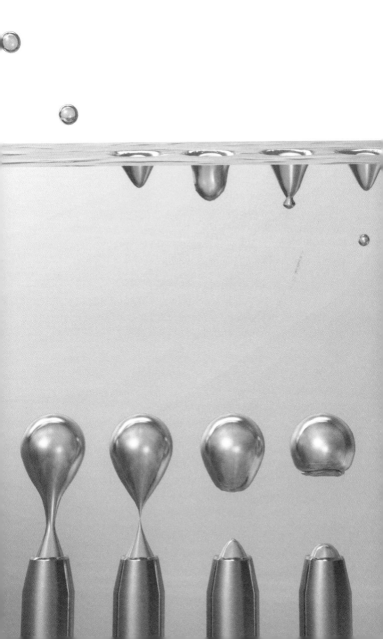

Hearing bubbles you can't see

Bubbles are beautiful acoustical objects because the gas inside them is very compressible. Water isn't – even in the deepest parts of the ocean, where the pressure is one thousand times atmospheric pressure, the water still has 95 per cent of its normal volume. In contrast, if you apply just one extra atmosphere of pressure to a gas at sea level, its volume will halve.

When you hear a bubble ring, it's because the bubble is pulsating: shrinking and growing ever so slightly so that the gas inside it compresses and then expands. The bubble walls push and pull on the water around them and this sends sound waves outwards.

One consequence of this is that bubbly water is effectively squishy, even if the bubbles are very tiny and take up only a tiny fraction of the total volume. And this affects the speed of any sound travelling through that water.

If you tap on the bottom of a mug filled with hot water, you'll hear a clear note. If you add instant coffee or hot chocolate granules and tap again, you'll hear a much deeper note, because the bubbles trapped in the granules are slowing the sound down. As you stir the coffee and the bubbles disappear, the note goes back up in pitch.

This effect is useful for detecting bubbles in pipes and tanks, where bubbles might not be seen but can still be heard. But this change in pitch just tells you that there are lots of tiny bubbles present. It doesn't give you many details. Exploring the water using other sounds can tell you much more.

Acoustic Christmas lights

On land, the best way to know what's going on around us is usually to look. But light doesn't travel far through water, so in the ocean it's often better to use sound. Humans, some whales, and dolphins use sonar to scan their surroundings, sending out a sound and listening for echoes coming back.

For any sonar user, bubbles twinkle like tiny festive lights. They're especially bright if the frequency of the sound matches the natural frequency of the bubble (the pitch it has when it rings like a bell). This is fantastically useful if you're an ocean bubble scientist like me – it lets us count bubbles of different sizes without disturbing them.

However, it's a nuisance if you're a ship trying to sneak past a submarine. When the navy first started using sonar from submarines, they quickly found that the bubbly wake behind a ship effectively drew an arrow on their sonar screens pointing at their target, because the bubbles scattered sound so strongly that their effect could be detected from miles away.

It's even more of a nuisance if you're a herring. Many fish have a swim bladder, a pouch filled with gas that helps them control their buoyancy. But this is effectively a bubble, so to a whale using sonar, dinner swims past lit up by its own acoustic candles. Sonar is also used by fishermen to locate shoals of fish very precisely. If you want to hide in the ocean, don't disguise yourself as a bubble.

So bubbles respond to sound very strongly. But what about the way that they interact with light?

Behold the bubble

It's worth asking why it's possible to see bubbles at all. Air is both transparent and colourless, and so is water. When you put one inside the other, shouldn't this little pocket of invisible gas inside an invisible liquid be lost from sight?

But it isn't. Even though water lets light through, it changes the direction of travel slightly. When light crosses the boundary between air and water, it swerves (physicists call this refraction). The greater the slant of the entry path, the greater the swerve at the boundary. And on its way out, the light swerves back the other way. So the path of the light ray has two kinks in it – one where it entered the bubble and one where it left. If the light hits at a very low angle, it may even bounce off the surface as though it was a mirror.

If you are looking at the bubble, what you see is that there's too much light coming from some places and too little from others. A spherical underwater bubble usually has a dark ring around the outside, because light from that region has either bounced off the bubble wall or has been deflected so that it comes out somewhere else. You still can't see the bubble itself, but you can see that something has shunted a patch of light around, and the bubble is guilty as charged.

Bubbles look beautiful almost by accident – the bubble itself is hidden, but it shapes the passing light so that its presence is revealed. And the effect of bubbles on light is most obvious when there are thousands or millions of bubbles together.

Why is foam always white?

When lots of bubbles form in a liquid but stay separate, sticking to each other instead of joining together, this new gas-liquid architecture is called a foam. Whipped cream, the head on beer and bath bubbles are all foams and they're almost always white, much to the disappointment of small children everywhere.

A foam is full of air-water boundaries, and light zigzags through the bubbly structure, getting refracted and reflected constantly as it goes. The liquid walls between the bubbles are very thin, so whatever dye might be in them won't have much of a chance to absorb any of the colours of the rainbow. Eventually, some light will find its way back out, and although it's been redirected many times, it hasn't been changed. So bath foam always looks white, because the light that goes in is the same as the light that comes out.

In a creamy milk foam, for example on top of hot chocolate, the protein surrounding the bubbles and the fat droplets in between help the foam hold more liquid, making it more opaque. There's enough liquid and fat to absorb some light, giving the foam a hint of chocolate colour.

It is said that Cleopatra took daily baths in asses' milk, to maintain the complexion of her skin. Had she had access to modern coloured bubble-bath solution to add to her baths full of milk she could have had vivid foamy baths with a beauty that rivalled her own. But foam is fun even if it isn't colourful, because all those bubbles stuck together have something that bubbly water doesn't have: structural strength.

Yummy bubbles

Foamy food is everywhere: mousse, ice cream, whipped cream, soufflé and the topping on lemon meringue pie. We love it because of the 'mouth feel' – the structure of the food is rebuilt around the bubbles, and this new architecture makes us think of creamy, rich indulgence. And that architecture is all based on the coating of each bubble: a thin layer of molecules or particles that forms a flexible cage for the gas inside.

If you lay a metal spoon flat across the top of a cappuccino, so that the bowl of the spoon is in the centre of the cup and the handle rests on the edge, the spoon will sit for several seconds before sinking through the foam layer. That's odd, because if you laid the spoon on either milk or air alone, the spoon would fall through immediately. But each tiny bubble of gas (perhaps 0.1 mm in diameter) is trapped within its own protein cage and, like a balloon, if you push on it, it'll push back. The combination of a liquid and a gas forms a structure that can behave like a solid, if you only push it gently.

The trick to all foamy food is to make tiny bubbles in a liquid that has the right mix of protein and/or fats to make those cages. In whisked egg whites the coatings are all protein, and in whipped cream they're all fat. But if the protein and fat are both present and compete with each other, the foam may well fall flat.

Food scientists are experts at making foams, because they're useful to convince our brains we're eating rich fatty food when we're not. And that bubble coat is important in helping the bubbles last longer too.

Long live the bubble!

It isn't just the bubbles in foams that have a coating. All soap bubbles do, and most individual underwater bubbles have at least a partial coating. It's unlikely that you've ever seen a perfectly clean bubble, one where the water directly touches the air everywhere on the surface. That's because lots of molecules have one end that tends to stay in water (hydrophilic) and one end that tends to stay out of water (hydrophobic). Fat and protein molecules come into this category, and so do soaps and detergents. As a bubble rises through water, it will sweep up any of these molecules it bumps into. Once stuck to the surface, both hydrophobic and hydrophilic ends are stable, so the molecule will stay put. And this coating helps the bubbles to last for longer before they pop or dissolve.

A completely clean bubble is squeezed by its own surface tension (which acts like the stretchy wall of a balloon), increasing the pressure inside the bubble and forcing the gases inside to dissolve into the watery walls. Left alone a perfectly clean bubble will squeeze itself out of existence in less than a minute. But a bubble coating provides stability, especially when it's got small particles embedded in it, by reducing the surface tension, preventing it from squeezing the bubble to nothing. Coated bubbles can last for hours, sometimes even days. And there's a lot you can do with a very long-lived bubble.

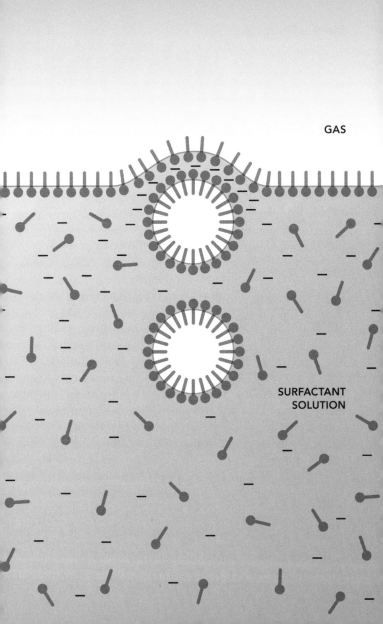

GAS

SURFACTANT
SOLUTION

Medical bubbles

Today, bubble coatings are being engineered to make very tiny bubbles so stable that they can be freeze-dried and kept in a pot in a cupboard. This is where bubble science meets medicine.

The first use of long-lasting bubbles in medicine was to make something called an ultrasound contrast agent. An ultrasound scanner is a type of sonar used to investigate the human body. Very high frequency sound is sent in, and by monitoring the echoes a doctor can build up a picture of what's inside. If very tiny bubbles (known as microbubbles) are injected into the bloodstream, they light up on an ultrasound scanner, just as the swim bladders of fish light up on sonar images. Doctors can follow the bubbles, tracking blood flow in real time and sharpening the detail on their scans.

But since all these tiny bubbles had to be coated to make them last long enough, it seemed like a good idea to put the coatings to work. So scientists have incorporated medicines into the bubble coating. If magnetic particles are also included, the microbubbles can be pulled towards a specific area of the body using external magnets. Standard ultrasound makes the bubbles oscillate gently, just enough to give their position away. But a focused burst of ultrasound can be targeted at one specific point, which kicks the bubbles at that spot into huge oscillations, forcing them to grow and shrink dramatically. This pops the bubble and releases the drug, exactly where it's needed. These treatments are still being developed but offer huge potential for the future.

But bubbles aren't just useful for our bodies. They're important for our planet.

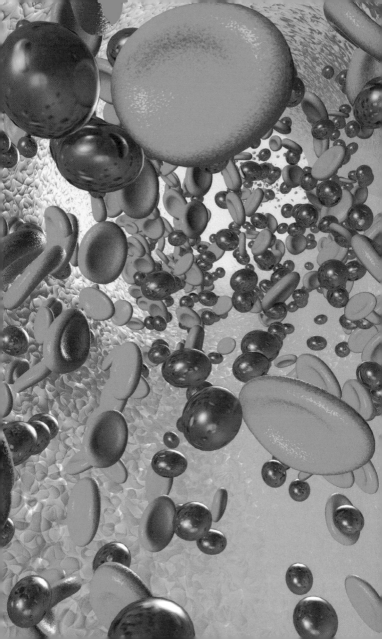

Bubbles in the ocean

Most of the bubbles on Earth are in the ocean, formed by breaking waves. We usually think of breaking waves at the coast, but most of them are far out at sea, where strong winds shove sideways on the ocean surface and produce waves. If the waves get steep enough, they will break, trapping air and forcing it down into the ocean. The patches of bubbles that come back up and sit at the surface are known as 'whitecaps', described by Rudyard Kipling as the 'mad white horses of the sea'. The surface bubbles are only a fraction of what the breaking wave produces – hidden beneath the surface there's a deep plume of underwater bubbles as well.

The same wave could break on a freshwater lake and on the ocean but only the ocean breaker will produce a visible whitecap. This is because the salt in seawater stops the bubbles joining together when they bump into each other – they bounce off instead. So the salty bubbles stay small and separate and can form a stable foam at the surface. In fresh water, the bubbles very quickly join together and these bigger bubbles burst as soon as they reach the surface, so no whitecap is formed.

Bubbles are also formed naturally where methane is escaping from the sea floor. Some patches of ocean, in places like the Santa Barbara Channel (which is close to the oil wells off the coast of California), are continuously fizzing, very gently, as tiny bubbles creep upwards.

These ocean bubbles don't just change the way the ocean looks. They also help the ocean breathe.

Travelling by bubble

At the ocean surface, bubbles act as vehicles that transport gases and particles between the atmosphere and the ocean. When a wave breaks, it carries tiny packets of the atmosphere down into the sea, where the air molecules may dissolve into the water. Our oceans have taken up about a quarter of all the extra carbon dioxide that humans have released from burning fossil fuels, which has reduced the amount of carbon dioxide in the atmosphere, and bubbles are important for this transfer process.

Bubbles also collect organic material as they rise through the ocean and back to the surface. The bubbles in a whitecap will be carrying a coating of carbohydrates, proteins, gel fragments, tiny particles, viruses and bacteria. When these bubbles pop, they spit liquid into the air in two ways. The first is that the 'cap' of the bubble, the thin curved film poking up above the surface, shatters into many tiny droplets which are carried off by the wind. The second is that the cavity left behind inverts and squirts a jet of water upwards, which can also split into droplets. The bubble coating is ejected upwards in these droplets, and when the water evaporates, tiny fragments of ocean goo are left floating on the wind. These tiny particles, called aerosols, drift upwards, affecting the way that light travels through the atmosphere and also how clouds form.

An individual bubble is minuscule compared with our planet, but because there are so many of them they're an important component in the ocean-atmosphere engine of Earth. And a different type of bubble can also act as a messenger from Earth's past.

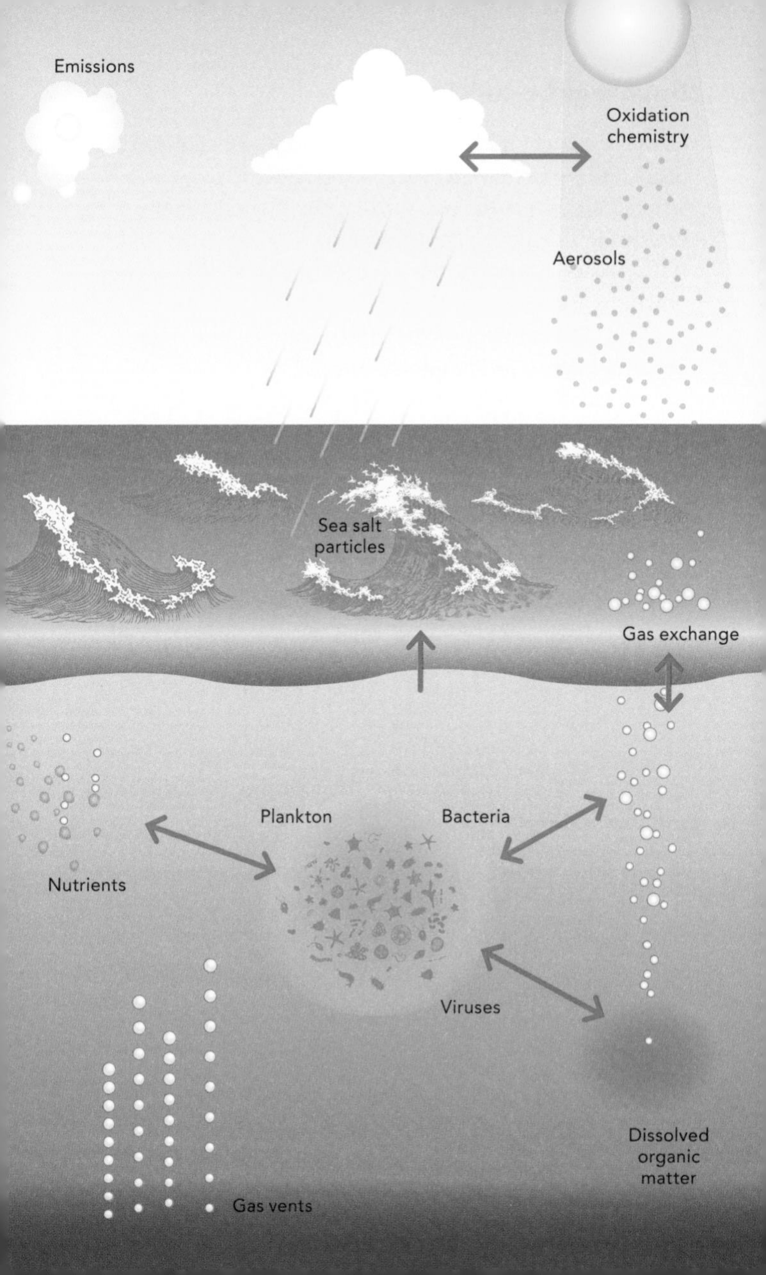

Emissions

Oxidation chemistry

Aerosols

Sea salt particles

Gas exchange

Plankton

Bacteria

Nutrients

Viruses

Gas vents

Dissolved organic matter

Bubble time capsules

Bubbles can act as vehicles in time, as well as in space. Snow falling on the great ice sheets of Antarctica and Greenland covers the previous layers of snow. As the snow builds up, the ice is compressed. Air can move through the fluffy layers at the top, but once a layer is 80–100 metres down the remaining air has all been squeezed into closed pockets. These tiny bubbles of gas, trapped in the ice, become atmospheric time capsules.

By drilling out long thin vertical columns of ice, known as ice cores, scientists can measure the past. The air inside the bubbles can be hundreds of thousands of years old, giving us a timeline of past climate.

If the scientists don't get there first, these trapped bubbles and their icy prison slowly flow out towards the coasts as glaciers. Many thousands of years later, they may break away from the glacier edge as icebergs. Up close to this floating icy rubble, you often hear a fizzing sound. The air in the ice bubbles is under huge pressure, and as the ice walls melt away in the ocean, each pocket of ancient atmosphere escapes with a dramatic 'pop!'. The fizzing sound is the bubbles of the past meeting the ocean of the present.

You'll find bubbles almost wherever you look in the natural world, unseen extras on the world stage, busy in the background. In some of those roles, they've been hiding in plain sight for years, without anyone suspecting their importance. Enter, stage left: the penguin.

Speedy penguins

The graceful and agile swimming of the Emperor penguin contrasts sharply with its laborious movement on land. But like all superheroes, its transition is partly hidden, in this case behind a cloak of bubbles.

To get out of the cold Antarctic Ocean, the penguins must swim upwards so rapidly that when they burst through the surface they have enough momentum to carry them up and on to the ice. There are leopard seals and other predators in the surface waters, so the faster the penguins swim, the better their chance of survival. Video footage of this crucial moment shows streams of bubbles covering the penguin's body and trailing out behind. And the bubbles help make these dramatic leaps possible.

Before going swimming, the penguins fluff up their feathers, trapping a store of air inside. Then they transform into aquatic acrobats, hunting 15 or 20 metres below the ice. Once they've caught their dinner, they start the trip back up to the surface. This is a sprint, and speed is everything. During the rapid dash upwards, they squeeze the feathers all over their bodies, releasing the trapped air and coating their bodies in bubbles. This bubbly coat reduces the drag of the water, enabling the penguin to swim upwards more than twice as fast as its normal cruising speed. It's an elegant solution to a very prosaic problem.

Engineers are working on applying the same principle to ships, to help improve fuel efficiency. It's a brilliant idea in theory. But ships don't have feathers, so there are still some practicalities to sort out before bubble coatings for ships become common.

The bubbles of the future

A bubble is a simple object, but physical riches spill
out of every detail of its nature. We have spent decades
understanding what bubbles do and why, and there is still
a vast amount to discover. But now we're starting to excel
at using bubbles as tools. They're not just playthings for
children – they're becoming some of the most important
building blocks of the modern world. Their potential will
only grow as we learn how to integrate their characteristics,
for example, combining their special surface behaviour and
their acoustical flexibility in the same application. Countless
new bubble-based technologies are under development, in
food, cleaning, medicine, energy generation and pretty much
any other field you can think of. Whether we can see and
hear them or not, the bubbles are coming.

But for all that technological wizardry, a bubble is still a
delightful toy for everyone, something to giggle at, pop,
play with and listen to. Bubble baths, fizzy drinks, streams
of soap bubbles and the cheery babbling of a stream are
here to stay. Appreciate the bubbles in your life: they may be
temporary, but they're fascinating, surprising, charming and
utterly beautiful.